Biomes
of the World

GRASSLAND

EDWARD R. RICCIUTI

BENCHMARK BOOKS

MARSHALL CAVENDISH
NEW YORK

Benchmark Books
Marshall Cavendish Corporation
99 White Plains Road
Tarrytown, New York 10591-9001

©Marshall Cavendish Corporation, 1996

Series created by Blackbirch Graphics, Inc.

· Printed in Hong Kong.

Library of Congress Cataloging-in-Publication Data

Ricciuti, Edward R.
    Grassland / Edward R. Ricciuti.
       p.   cm. — (Biomes of the world)
    Includes bibliographical references (p.   ) and index.
    Summary: Describes the world's vast expanses of grassland, how they developed, and the plants and animals that live there.
    ISBN 0-7614-0136-9 (lib. bdg.)
    1. Grassland ecology—Juvenile literature. 2. Grasslands—Juvenile literature. [1. Grasslands. 2. Grassland ecology. 3. Ecology.] I. Title. II. Series.
QH541.5.P7R48   1996
574.5'2643—dc20
                                     95-41072
                                           CIP
                                           AC

Contents

Introduction

People traveling in an airplane often marvel at the patchwork patterns they see as they look down on the land. Fields, forests, grasslands, and deserts, each with its own identifiable color and texture, form a crazy quilt of varying designs. Ecologists—scientists who study the relationship between living things and their environment—have also observed the repeating patterns of life that appear across the surface of the earth. They have named these geographical areas biomes. A biome is defined by certain environmental conditions and by the plants and animals that have adapted to these conditions.

The map identifies the earth's biomes and shows their placement across the continents. Most of the biomes are on land. They include the tropical rainforest, temperate forest, grassland, tundra, taiga, chaparral, and desert. Each has a unique climate, including yearly patterns of temperature, rainfall, and sunlight, as well as certain kinds of soil. In addition to the land biomes, the oceans of the world make up a single biome, which is defined by its salt-water environment.

Looking at biomes helps us understand the interconnections between our planet and the living things that inhabit it. For example, the tilt of the earth on its axis and wind patterns both help to determine the climate of any particular biome.

The climate, in turn, has a great impact on the types of plants that can flourish, or even survive, in an area. That plant life influences the composition and stability of the soil. And the soil, in turn, influences which plants will thrive. These interconnections continue in every aspect of nature. While some animals eat plants, others use plants for shelter or concealment. And the types of plants that grow in a biome directly influence the species of animals that live there. Some of the animals help pollinate plants. Many of them enrich the soil with their waste.

Within each biome, the interplay of climatic conditions, plants, and animals defines a broad pattern of life. All of these interactions make the plants and animals of a biome interdependent and create a delicate natural balance. Recognizing these different relationships and how they shape the natural world enables us to appreciate the complexity of life on Earth and the beauty of the biomes of which we are a part.

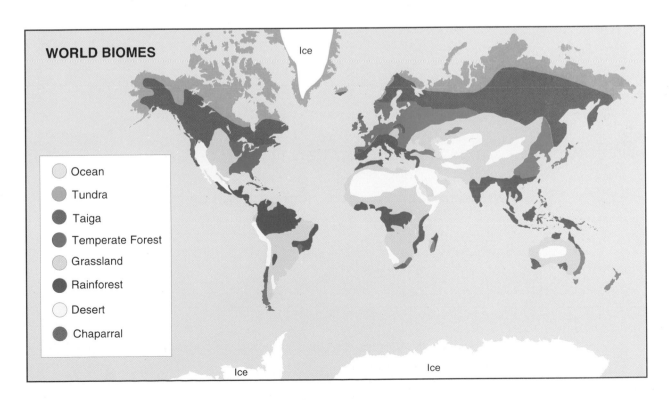

WORLD BIOMES

- Ocean
- Tundra
- Taiga
- Temperate Forest
- Grassland
- Rainforest
- Desert
- Chaparral

1

A World of Grass

Stirred up by millions of hooves, dust billows into the air, which is filled with the sounds of braying, snorting animals. Tossing their horned heads, their tails switching back and forth, over a million wildebeests are on the move. They are migrating across the great Serengeti Plain of Tanzania, one of Earth's vast grasslands.

The wildebeest herd of the Serengeti is one of the greatest concentrations of large wild mammals left on the planet. The wildebeests, a species of African antelope, spend the year following the rainy season from north to south and then back again. With them go many other creatures, such as zebras, gazelles, and buffalo—and the animals that prey on them, such as lions, cheetahs, and wild dogs. Following an age-old, endless cycle, the wildebeests chase the rains in search of the water and grass that enable them to survive.

Opposite:
A large herd of wildebeest graze on the Serengeti Plain.

7

The migration across the Serengeti is one of the most thrilling wildlife spectacles on Earth. Perhaps nowhere else can so many large creatures be seen on the move together.

A Numbers Game

Scientists studying the relationships between animals and their environment often compare grasslands and rainforests. On a grassland, grasses far outnumber any other type of plant life. Because of this, grasslands support fewer species, but greater numbers, of animals than rainforests do. In the rainforests, vegetation is extremely varied, so this biome can support many species of smaller populations than grasslands.

In healthy grasslands, African elephants can find the food they need.

For those animals that can subsist on grasses—from seed-eating birds and termites to pronghorn antelopes—a grassland provides an almost endless supply of food. These creatures, like the wildebeests of the Serengeti, exist in huge numbers.

At one time, the grasslands of North America supported 60 million American bison. This was perhaps the greatest concentration of a large-animal species ever. Many grassland animals are massive; the biggest living land creature is the African savanna elephant. Because trees are scattered on the grasslands, a large body does not impede an animal's movement, as it would in the forests.

Grasslands of the World

Grasslands cover more than 14 million square miles (36 million square kilometers), more than a quarter of the Earth's surface. The greatest expanses of grassland are found in the western United States, Central Asia and adjacent Europe, and Africa. Australia also has large grasslands on the rim of its central desert.

Not all of the grasslands are the same. There are several types, ranging from those virtually without trees to those with scattered woodlands. Treeless grasslands include the prairies of North America, the pampas of South America, and

The great plateau of the altiplano stretches out beneath the Andes in Peru.

THE BADLANDS

In South Dakota lies a prairie landscape called the badlands. The name badlands was first applied to this area by the Sioux, who called it *mako sica*, meaning "bad land." There are scattered areas of badlands in North Dakota, Wyoming, Colorado, Kansas, Nebraska, Arizona, California, and Alberta, Canada. But the largest area has been preserved in South Dakota's Badlands National Park.

The badlands are a beautiful but sometimes spooky place. One might imagine a landscape on an alien world looking like this. The grassland is carved into mazes of canyons. Rocky spires, ridges, and buttes rise into the sky. Enormous rocks, from which soil has been eroded, are banded with hues of red, brown, and orange.

More than 35 million years ago, the region that is now the badlands was a swampy, forested plain. Many prehistoric mammals, such as saber-toothed tigers and titanotheres, inhabited the plain. Titanotheres were horned animals that resembled rhinoceros. As time passed, however, the climate cooled and became drier. The swamp forests vanished and eventually were replaced by grasslands.

The Ice Age glaciers never reached the area, but the climatic conditions that created the ice sheets helped its distinctive landscape to form. Wind and rain eroded the soil, as did streams flowing from the foot of the glaciers to the north. Eventually, this erosion created the landforms seen in the badlands today. The landscape may once again look different in the future because the process of erosion by wind and water will continue to shape the land.

These beautiful, cone-shaped land formations are preserved in South Dakota's Badlands National Park.

A GRASSLANDS PARTNERSHIP

Before the grasslands of the American West were turned largely into cattle pastures and farms, huge herds of bison and great colonies of prairie dogs played a major role in maintaining them. The prairie dogs were even more numerous than the bison. One prairie dog colony, or town, examined in 1900 in Texas was found to contain 400 million prairie dogs and covered 25,000 square miles (64,750 square kilometers).

Prairie dogs feed on grasses. One animal can eat up to 2 pounds (0.9 kilogram) a day. However, a thick growth of grass can also prevent them from entering an area. They must work hard to cut down the grass in order to see approaching enemies. Heavy growths of grass also interfere with the digging of burrows, in which prairie dogs live. When huge herds of bison moved through the thick grass, they grazed and trampled it down, opening up the territory for the prairie dogs.

Once this happened, the prairie dogs were able to move in and dig burrows. The soil over which the bison had traveled was compacted. Little air could penetrate it, and the roots of grasses had a difficult time spreading. The burrowing prairie dogs loosened the soil, and the bison left droppings that became fertilizer. Thus, both of these animals promoted the growth of grass.

the steppes of Europe and Asia. In some places, grasslands can be found high in the mountains. Below the peaks of the central Andes Mountains, at an elevation of about 11,000 feet (3,355 meters), is a great plateau called the altiplano, or high plain. Cold and windswept, it is covered with grasses and tough, grasslike plants. It is barren of trees.

When they are near deserts, grasslands may be covered with low bushes, many of them thorny, called scrub. Grasslands with scattered trees, which are more extensive in Africa, are called savannas.

How Grasslands Developed

Whether a grassland can exist in a given area is determined mostly by the amount of rain that falls in the area. If there is less than 10 inches (25 centimeters) of precipitation a year, a desert eventually results. If rainfall is at least 40 inches

(101 centimeters) annually, a forest grows. Grasslands occur where amounts of rainfall are somewhere in between, although some grasslands exist in places that would also support forests.

For millions of years, long-term, global climatic changes have caused "wars" between the grasslands and the forests. During periods of wet weather, the forests have expanded and marched into the grasslands. During dry spells, the forests have retreated before the advancing grasses. At present, the savanna covers a much greater portion of Africa than the rainforest does. Millions of years ago, however, Africa underwent wet periods called pluvials. African rainforests occupied more territory during the pluvials than they do today.

Before the Rocky Mountains arose, about 65 million years ago, the area now occupied by the North American prairie was largely forested. The forests were watered by moisture-laden winds that blew across the continent from the west. When the winds crossed the newly formed Rockies, they dropped their moisture. East of the Rockies, the climate became too dry to support trees, and grasslands developed.

Winds also allow the grasslands to thrive. Evaporation occurs more rapidly when the wind is blowing, which is why a breeze cools your skin on a hot summer's day. Strong, steady winds regularly blow over the grasslands, reducing the amount of moisture that trees need in order to grow.

Natural fires, which are started by lightning, also help the grasses win their battle against the trees. So do fires set by people—scientists believe that burning by Native Americans greatly aided the growth of prairie grasses. Other plants that compete with the grasses are also destroyed by fire. Grass plants survive because about 70 percent of their bulk is underground, in the form of roots and rhizomes, underground stems that give off new shoots. This underground mass is not affected by flames.

Moreover, when a fire burns a grassland, it destroys the litter of dead plant material that, were it not removed, could blanket and kill the grasses by blocking out essential sunlight. (Lawns die, for example, when cut grass is allowed to accumulate over a long period of time.) The more light that grass leaves receive, the faster and larger their roots and rhizomes grow. These parts of a grass plant are in contact with the soil, from which they absorb nutrients and moisture. The more surface the roots and rhizomes have, the healthier the plant is. Studies have shown that the amount of material in the roots and rhizomes of grass in lands that have been regularly burned is almost 40 percent greater than the amount in grasses in unburned areas.

Another force that often helps the grasses thrive is the large numbers of grazing animals that trample and kill tree seedlings, eliminating the competition. However, if too many

This savanna in Venezuela is burned yearly to activate new growth.

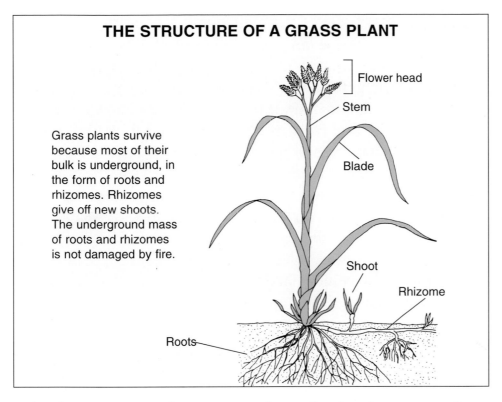

THE STRUCTURE OF A GRASS PLANT

Flower head

Stem

Blade

Grass plants survive because most of their bulk is underground, in the form of roots and rhizomes. Rhizomes give off new shoots. The underground mass of roots and rhizomes is not damaged by fire.

Shoot

Rhizome

Roots

animals concentrate in one area of grassland and overgraze it, the grasses will die.

In Africa, elephants feeds on the leaves and the bark of trees, as well as on the grass. When an elephant eats its way through a grove of trees, it leaves little behind. This also helps the African grasses to compete with the trees.

Adaptations for Grasslands

Grasses are tougher than many other kinds of vegetation, such as leaves. Animals that depend on grasses for food usually have teeth that are suited to cutting and grinding tough plant material. Animals like zebras and antelope have wide, chisel-shaped incisors, or front teeth, that clip off grasses, and broad molar teeth toward the rear of their mouths for chewing them. The lips of horses are prehensile, or grasping, while those of cattle are broad and strong. Both are useful for picking up grass. Horses have digestive juices especially suited to hard-to-digest grasses. Cattle and antelope have stomachs with four chambers. These animals swallow partially chewed grass, which enters the first two chambers, where bacteria chemically break down the grass into a soft mass. Then they regurgitate the mass, or cud, and chew it further before swallowing it. It then goes to the first two chambers again and then to the other two stomach chambers, where digestion is completed.

Many grassland mammals have long legs and can run with considerable speed. Hooves are a big advantage for an animal that must run over the hard surface of a grassland because they protect the feet. The single hoof of the wild horses, such as zebras, offers the most protection.

It is not by coincidence that the cheetah, which inhabits the grassland and can run up to 60 miles (97 kilometers) an

A lion (bottom left) charges a group of zebra, wildebeest, and springbok in a kill attempt. Grazing animals rely on speed to escape their predators.

hour, is the only cat that cannot withdraw its claws. The cheetah's claws are blunt and thick, like those of a dog, and adapted for running. Cheetahs catch small antelope by chasing them down. Lions, which have claws similar to those of the average cat, capture prey in the usual feline way. They creep close to their target and then rush it.

Grasslands soil is fine and often deep. Many of the small creatures of this biome, such as lizards and rodents, live in burrows. The large crop of seeds produced by the grasses feeds rodents, birds, and many other animals.

Rodents, lizards, and other small creatures provide a bounty of food for hawks and eagles, so these hunters of the sky are a frequent sight above the grasslands. The wide-open landscape of the grasslands makes hunting easy for these birds of prey, which are able to cover great distances and survey the ground below with keen eyes.

Sooner or later, all animals and plants of the grasslands die. Their remains are then broken down by bacteria and other organisms and are returned as nutrients to the soil, where they nourish the grasses that are the very source of life in the grasslands biome.

Eagles are a common bird of prey on many grasslands.

17

2

A Carpet of Grass

The number of plant species that can grow in a grassland depends on how much moisture it receives. Grasslands that receive more than 20 inches (51 centimeters) of rain yearly may have more than 200 types of flowering plants, while those that verge on being deserts may contain fewer than 100 types. Rain is the greatest shaper of grasslands. Most rain that falls on grasslands in temperate regions does so in the spring. Rainy seasons in the grasslands of tropical regions vary.

Grasses, Tall and Short

Varying amounts of rainfall have created three different types of prairie in North America. The easternmost is the tall-grass prairie, most of which has been converted to farmland and other types of development, such as towns, cities, and mining areas. The typical grasses of each kind of prairie include

**Opposite:
This grassland community in the National Bison Range in Montana is a lush prairie habitat.**

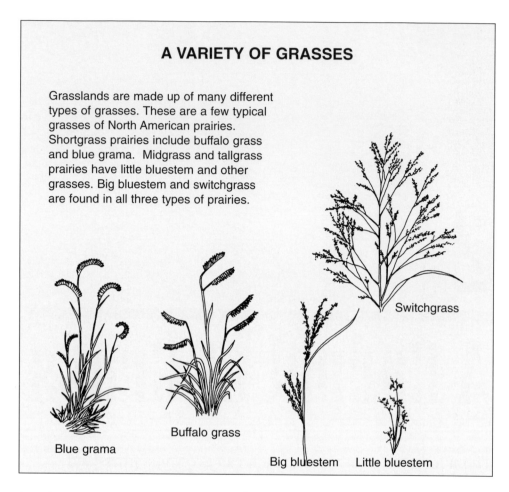

A VARIETY OF GRASSES

Grasslands are made up of many different types of grasses. These are a few typical grasses of North American prairies. Shortgrass prairies include buffalo grass and blue grama. Midgrass and tallgrass prairies have little bluestem and other grasses. Big bluestem and switchgrass are found in all three types of prairies.

Switchgrass

Blue grama

Buffalo grass

Big bluestem   Little bluestem

big bluestem and switchgrass, which can grow 7 feet (2 meters) tall. Tallgrass prairie grows where rainfall is about 30 inches (76 centimeters) a year, sometimes even more. This is the place where the competition between the grass and the trees is the fiercest.

To the west, where rainfall is between 20 and 30 inches (51 and 76 centimeters) a year, the tallgrass prairie blends into the midgrass prairie. There, such grasses as little bluestem and needlegrass grow to about 4 feet (1.2 meters) high. The westernmost prairies, which reach the Rocky Mountains, are the shortgrass prairies, where buffalo grass and blue grama grow between 4 and 18 inches (10 and 46 centimeters) tall. The shortgrass prairie is sometimes called the High Plains.

Similar relationships between rainfall and grass growth exist on grasslands elsewhere in the world. The southern portion of the Serengeti Plain is the driest, and it is there that the shortgrass plains are found. In the northwestern part of the Serengeti, which gets more rainfall, are the tallgrass plains. During the wet season, in the southern Serengeti, the flat plain is covered with a blanket of low grass that resembles a green carpet. In the north—even during the dry season—one can sit in a four-wheel-drive vehicle amid brown grasses that grow as high as its roof and can hide animals as big as a rhinoceros.

An African elephant is almost completely hidden by tall grass in Tanzania.

The grassland is a world between desert and forest. Near the border of North America's eastern forests, the moist tallgrass prairie is studded with scattered trees. In the west, the shortgrass plains become increasingly arid until, eventually, the grasses thin out and are replaced by desert vegetation. The steppes of Russia are similar in character. To the south, they are bordered by deserts. The steppes just north of the desert are dry and dotted with shortgrasses. North of those are the tallgrass steppes, where the grasses can grow taller than 6 feet (1.8 meters). Still farther north are steppes with scattered trees that are similar to savannas. The northern borders of these steppes merge with forest.

Scattered trees, such as this flat top acacia, are found throughout the grasslands of Africa.

GRASSES IN FLOWER

Most grasses flower in the early summer in temperate regions and after they have been soaked during the rainy season in tropical areas. The stems that bear flowers stand straight up and are fast-growing. They bear a few leaves and are topped by spike-shaped flower clusters. Grass flowers are dull-colored and very simple. They have no fragrance and no petals. They consist only of the stamen, the male part of the flower, and the carpel, the female part. The stamen and the carpel of most grass flowers are tiny and difficult to see.

When the carpel is ripe, it produces pollen, which is carried on the wind. Atop each carpel is a sticky structure called a stigma. Pollen that lands on the stigma sticks to it and fertilizes the ovule, the female sex cell in the carpel. The ovule then grows into a seed that, if conditions are right, will produce a new grass plant.

Tiny flowers of muted color bloom on a blade of blue grama grass.

In Africa, the trees that are sprinkled across the savanna, such as acacias, are least abundant near desert regions. On dry savannas, acacias grow near riverbeds. They dot the savanna in places where rainfall is heavier. African savannas occur both north and south of the rainforest that occupies the center of the continent. The northernmost area of savanna,

23

sandwiched between the forest and the Sahara Desert, lies in a belt that extends from coast to coast.

Savannas also exist in South America and Australia. In South America, they lie to the north and south of the Amazon rainforest. Many large rivers wind through the South American savanna. During the rainy season, these rivers overflow, flooding vast areas. The areas of Australian savanna are on the edges of that continent's bone-dry interior. Most Australian savannas are very arid, almost desertlike in character, although some in the north are well watered.

Green Grow the Grasses

There exist more than 7,500 species of grasses, and they range in size from about 1 inch (2.5 centimeters) high to 120 feet (37 meters) tall, in the case of bamboo, which is the largest of all the grasses. Rice, corn, oats, and other cereals are also grasses that have developed from wild varieties.

A grass has a round, hollow stem and long, narrow leaves, which are called blades. Grass blades grow from the base, so that when their tops are cut, they can regenerate. Grasses reproduce in three ways. They have flowers whose pollen is spread by the wind. When the grass flower is pollinated, it produces a tiny seed that grows into a new grass plant. Grasses also reproduce by sending out rhizomes, along which buds develop and grow into plants. Some grasses also produce new shoots from the base of their stems.

During spring and early summer, temperate grasslands are ablaze with the flowers of plants, such as asters, that live among the grasses. The steppes of Russia are brightened by red tulips, peonies, anemones, irises, and sage. Some of the grasses themselves are beautiful. In the spring, feather grasses also develop plumes of tiny flowers that wave in the wind.

The American prairies burst into springtime color with the blooming of such wildflowers as prairie smokes, Indian paintbrushes, blazing stars, coneflowers, strawberries, phlox,

crocuses, gentians, and violets. Wildflowers are especially abundant on the tallgrass prairie.

Africa's grasslands are not as rich in wildflowers as those of the temperate climates are. During the dry season, the grasses turn brown and stop growing. When the rainy season arrives, new grass growth turns the landscape a brilliant green. The transformation takes place in only a few days.

A field of Indian paintbrushes color a Texas prairie in the springtime.

UNDERGROUND RICHES

The grassland's topsoil is tremendously rich in nutrients and often several feet in depth. This soil was formed millions of years ago, when rock particles that were eroded from mountain ranges were swept onto the plains by wind and water. The process continues today as rocks break down into the tiny mineral particles that are among the building blocks of soil. The grasses themselves also help to create the soil. As grasses and other organisms die, their remains are broken down and decomposed by bacteria, fungi, worms, and other residents of the soil. The resulting organic materials mix with rock particles and clay to form a rich mix of nutrients that plants, including grasses, need in order to grow.

The grasses' roots—some of which can extend 6 feet (1.8 meters) deep into the earth—hold the soil together and prevent it from being washed or blown away. Soil particles in the grassland are very fine. This enables roots to penetrate the soil easily. If the grasses are destroyed, however, no roots remain to hold the fine-textured soil in place. It can then be easily eroded.

Drought Fighters

Periods of drought are not uncommon in the grasslands, and many grasses have adaptations that enable them to live through dry spells. Some grow very quickly during the rainy season, accomplishing the job of gaining size and producing seeds before the climate becomes dry. Then they all but stop growing until the rains come again. The grasses with very deep roots survive dry periods because they can draw on moisture that remains far below the surface. The leaves of some grasses curl up during drought, reducing the amount of surface area from which they can lose water through evaporation. Many grasses have hairs on their roots that soak up water.

Alike but Different

Grasslands around the world have several things in common. However, they differ in some ways as well: They support different species of grasses and other plants; and they are home to different types of animals that have evolved in various parts of the world.

The grasslands of North America were—and in a few places still are—populated by great bison herds, pronghorn antelope (although the pronghorn is called an antelope, it belongs to a family entirely its own), mule deer, coyotes, and wolves. The African grasslands are inhabited by scores of antelope species, as well as buffalo, elephants, lions, leopards, and cheetahs. Vast herds of saiga antelope, wild asses, and wolves roam the grasslands of Eurasia, and kangaroos are the major grass eaters in Australia. The South American savannas support capybaras, different species of deer, and jaguars. Because the grasslands are so wide-open, many of the animals that inhabit them can easily be seen. No better place exists on Earth for watching wildlife.

The variety of animals that depend on grasslands for their survival is vast. Here, Indian wild asses travel in a herd.

3

On and Above the Grasslands

During the rainy season, which lasts from November to May, the great herd of wildebeests in Serengeti National Park feeds on the green shortgrasses in the southeastern part of the region, in and around the Ngorongoro Conservation Area. Their dark forms dot the plain as far as the eye can see. In January and February, the pregnant females give birth to calves, which are born at a time when food is plentiful. By May, the rains are ending, and the wildebeests set out on their journey northward.

**Opposite:
A group of
vultures
surveys the
Masai Mara in
search of food.**

Thomson's gazelles move rapidly across the Serengeti.

The Thundering Herds

Long lines of wildebeests, marching nose to tail, travel over ancient game trails toward the northwestern portion of the park. They stir up the land, and they surge across the rivers. The weak among them are likely to die on the way and become food for predators, such as lions, and scavengers, such as vultures.

The plain they have left behind is turning brown, and water has become scarce. In the northwest, however, the rains are still falling. They keep the grasses green and provide water for drinking. As the rains move even farther north, the wildebeests continue their journey. They move across the border into the Masai Mara of Kenya, which is a vast area of scattered woodlands and grasslands. Toward the beginning of November, the herd is on the march again, now following the rains as they move south. By the time the wildebeests reach the Ngorongoro region, the grass is green, ready to be eaten.

Accompanying the wildebeests are huge numbers of other animals, including 275,000 Burchell's zebras, 1,500 elands, 725,000 Thomson's gazelles, and 52,000 Grant's gazelles. The wildebeests—and the animals that travel with them—have developed their adaptation to the regular patterns of wet and dry seasons over a vast amount of time. Fossils found in Tanzania that date to a million years ago indicate that the same species of wildebeest grazed in the Serengeti during some parts of the year, but not during others, suggesting that even that long ago they were migrating into and out of the area.

Picky Eaters

How is it possible that 1.5 million wildebeests and almost a million other grazing animals do not consume all of the grass? One explanation is that they travel elsewhere during the dry season. Another is that while most of these animals will consume whatever grass is available when food is in short supply, each species will eat only the type of grass it most prefers when it is plentiful.

Zebras, like wild horses, have incisor teeth in both upper and lower jaws, unlike antelope, which lack upper front teeth. Zebras graze down the taller, tougher parts of the grass. Wildebeests then can feed on lower, more nutritious parts. After that, gazelles can reach the tender shoots that remain.

A similar relationship existed in the American West between the bison and the pronghorn. Today things have changed. The vast bison herds are gone, but the pronghorn

Zebras and wildebeests share grazing areas.

remain, in considerable numbers. Before the settlement of the West, however, the western plains and prairies—especially the midgrass prairie—were shared to a large extent by both species. Even so, the bison preferred moister areas, where the grass was longer, and the pronghorn gravitated to shortgrass plains in dry areas. (The bison is primarily a grass eater, while the pronghorn eats other small plants and even cacti.) By consuming different foods, they did not exhaust any one source of food.

Running into the Wind

The saiga antelope, which inhabits the steppes of Eurasia, is the only antelope found within Europe. There are more than 2 million saigas living in the wild. Much of the region that they inhabit can be very cold and snowy in winter. When the heavy winter storms arrive, the saigas may be temporarily cut off from the grasses on which they feed. They flee the storm by literally running for their lives into the wind, going in the opposite direction as the approaching bad weather. The saigas also migrate between summer and winter grazing areas, sometimes covering several hundred miles. They even migrate from north to south during severe winter weather in search of areas where the snow cover is light.

The saigas, like many other antelope, have a number of adaptations that help to protect them from danger on the grasslands. They have excellent long-distance vision—an advantage in wide-open spaces—and can spot an enemy more than 3,000 feet (915 meters) away. Young saigas can run swiftly at a day or two of age, and within another few days they are eating plants as well as drinking their mothers' milk. Before they are three months old, they are feeding entirely on plants, which enables them to be more independent. Saigas are related to gazelles, a number of which live on the Asian grasslands.

Antelopes, such as this Russian saiga, are well adapted to life in grasslands.

33

Giant Rodents and Small Camels

The South American grasslands are home to the world's largest rodent and to small members of the camel family. The rodent is the capybara, which lives in well-watered tropical grasslands where streams seasonally flood large areas. Weighing up to 150 pounds (68 kilograms), the capybara is an excellent swimmer. It often takes to the water to avoid terrestrial enemies.

The South American wild camels are the guanaco and the vicuña. Unlike their larger relatives, they lack humps. Guanacos live at sea level and up to elevations of 18,000 feet (5,490 meters) on grassland plateaus in the mountains. Vicuñas inhabit only the high plateaus, such as the altiplano. Much of the land inhabited by the guanacos and the vicuñas has a harsh climate. Both species have broad incisors for cutting the tough grasses and other low plants that grow there.

Vicuñas graze on the tough grasses and low plants of the altiplano.

Pouched Grazers

Kangaroos of several species—particularly the largest ones—are the big grazing animals of Australia. Like other grazers, these pouched mammals have incisors that are suited to cutting off grass. Kangaroos are marsupials, an order of mammals in which the females have a pouch to carry their young. One species, the grey kangaroo, bears its young just after the rainy season, when grass growth is heaviest. Kangaroos are known for their speed and ability to travel in long bounds.

Hunters of the Grasslands

Predators in the grasslands use a few basic tactics in order to capture their prey. The cheetah rockets after small antelope, twisting and turning with its victim and relying on speed. Most other cats, such as the cougar and the caracal, hide in the grass, creeping as close as possible to their prey, then launching a sudden, overwhelming attack. Lions use this method too. However, because they live in family groups, called prides, they sometimes will join forces to surround prey or even ambush it.

African wild dogs and wolves generally rely on cooperation when they hunt. They often pursue their prey over long distances until they tire it out or wound it so severely that it can no longer flee. Coyotes sometimes chase deer in a group, but often hunt smaller creatures on their own or in pairs. Dingoes, Australian wild dogs, are also solitary hunters. One of the strangest members of the dog family is the so-called

A gray kangaroo carries her baby (called a joey) in her pouch.

Cheetahs use their great speed to capture prey. Here, a small Thomson's gazelle is caught in the Serengeti.

maned wolf of the South American grasslands. It has extremely long legs and a small head, and it resembles a fox on stilts. It hunts alone, chiefly for small animals, and also feeds on fruit. Its extra-long legs are probably an adaptation for helping the animal to see above the grass.

The Clean-Up Crew

Hyenas, too, can be fierce predators, attacking antelope and other large creatures in snarling packs. They are also scavengers, eating the remains of animals that other predators have left behind.

The role of scavenger is very important, and the grasslands have many of them. Jackals, small members of the dog family, often feed on the kills of larger animals. Most of the scavengers, however, are birds. An old lion kill on the African grassland usually attracts hordes of vultures and long-legged marabou storks. Scavengers are the last link in a chain in which energy is passed from one organism to another in the form of food. It begins when grass uses sunlight to grow, then is eaten by grazers, which in turn are fed upon by predators. In the end, the scavengers eat what is left—of animals that have been killed and of those that have died of other causes.

A GRASSLANDS FOOD CHAIN

1. The grasses use sunlight to grow.

2. The grasses are eaten by herbivores, such as gazelles.

3. Carnivores prey upon herbivores and other animals.

4. Scavengers, such as hyenas, feed on the carcasses of animals that have been killed by predators or have died from other causes.

BIG BIRDS

They are giants among living birds. Only a few prehistoric birds were larger. They look alike, with long necks, wings that cannot take them aloft, and long, powerful legs for running. All are animals of the grasslands, yet they live oceans apart and are not closely related. The ostrich inhabits Africa; the rhea, South America; and the emu, Australia. Each has evolved adaptations that suit it to life on the wide-open plains and prairies. These adaptations have made them similar in behavior and appearance.

The ostrich is the biggest living bird at almost 350 pounds (159 kilograms) and 8 feet (2.4 meters) tall. Ostriches live in groups of about 12 to 50 animals. A male may mate with up to five females, all of which lay their eggs in the same nest, which the male incubates.

Rheas weigh up to 55 pounds (25 kilograms) and are 5 feet (1.5 meters) tall at most. Their groups number about two dozen individuals. Like ostriches, one male may mate with several females and incubate eggs in the same nest. Emus can grow to 6 feet (1.8 meters) tall and weigh up to 100 pounds (45.4 kilograms). They also live in small groups, and, as do the other big birds, the male incubates the eggs.

Even though ostriches have large wings, they cannot fly. Here, a female ostrich displays her feathers.

Above the Grasses

Vultures continually soar above the grasslands, searching for food with their keen eyes. Because the grasslands are suited to birds that fly long distances and use their eyes to spot food from a great height, they are home to a great number of hawks and eagles.

Birds of prey hunt from the air. In Africa, however, they have a close relative that hunts on land. The long-legged secretary bird—so named because plumes on its head resemble the feather pens that clerks once kept behind their ears—chases after snakes, lizards, and other small mammals. It takes great strides, seizing its prey with its feet, which have talons similar to those of a hawk or an eagle.

The grasslands, with their abundance of grass seeds and insects provide a wide variety of food for birds. The grasslands of western North America, for example, are inhabited by many species of sparrows, meadowlarks, prairie chickens, swallows, flycatchers, and warblers. As one might expect, many grassland birds nest on the ground. Others nest in riverbanks, canyons, or trees. Many bird species use grasses to line or to build their nests—another way in which grasses contribute to the life of the biome that is named after them.

The secretary bird's long legs and sharp claws make this bird of prey a successful hunter on the ground.

4

Under the Grass

There are some places in western North America where one can still see large prairie-dog towns. One of these is the Badlands National Park of South Dakota. Mounds of dirt above prairie-dog burrows dot the flatlands between the canyons, spires, and buttes. Prairie dogs seem to be every-where: sitting atop burrows, on the lookout for enemies, feeding on plants, and scurrying about. Prairie dogs are among the many animals, mostly rodents, that burrow beneath the surface of grasslands around the world.

Opposite:
A black-tailed prairie dog looks out from its burrow in North Dakota.

A Home Underground

Prairie dogs belong to the squirrel family and, like many other types of squirrels, live on the ground. They are small animals that weigh only 2 to 3 pounds (0.9 to 1.4 kilograms).

Among the sounds they make are the barklike noises that gave them their name. There are two kinds of prairie dogs. The black-tailed prairie dog is the most numerous and widespread and usually inhabits midgrass plains. The white-tailed prairie dog is found most often on meadows in the mountains.

A prairie-dog family, which is headed by a dominant male, makes its home in a burrow, where the animals spend about half their time. Up to 30 feet (9 meters) long, a prairie-dog burrow has many tunnels and chambers. There is a listening chamber, a dead-end compartment where the prairie dogs wait and listen for enemies before emerging aboveground, a nesting chamber, food-storage chambers, a "bathroom," and sleeping chambers. At one end of the burrow is an escape hatch, lightly covered with earth, through which the prairie dogs leave if an enemy enters their underground home. The entrance to the burrow is surrounded by a circular heap of dirt, which keeps water out during heavy rains.

Life in Town

Prairie-dog towns are not merely concentrations of family units. Instead, there exists an organization of sorts, rather like a human town. The family groups of black-tailed prairie dogs are called coteries, and they have their own home territories. Usually, the coteries include about eight animals: an adult male, three or four adult females, and a number of young. Young males leave the coterie territory as they begin to mature and try to establish coteries of their own. The members of a coterie will fight to keep outsiders away from their territories. Because prairie dogs generally respect the boundaries of their neighbors, however, battles are infrequent.

Members of a coterie learn to know one another by kissing, wagging their tails, and combing one another's fur to remove dirt, debris, and parasites. The more these activities go on, the closer the bonds between the animals become. If one prairie dog does not respond to the gestures of another, it

A PRAIRIE-DOG BURROW

Prairie dogs build elaborate burrows with many chambers that are built off of a main tunnel. There are sleeping chambers, grass-lined nesting chambers, toilet chambers, and listening chambers. Prairie dogs are very organized animals and constantly adapt their burrows to fit their needs.

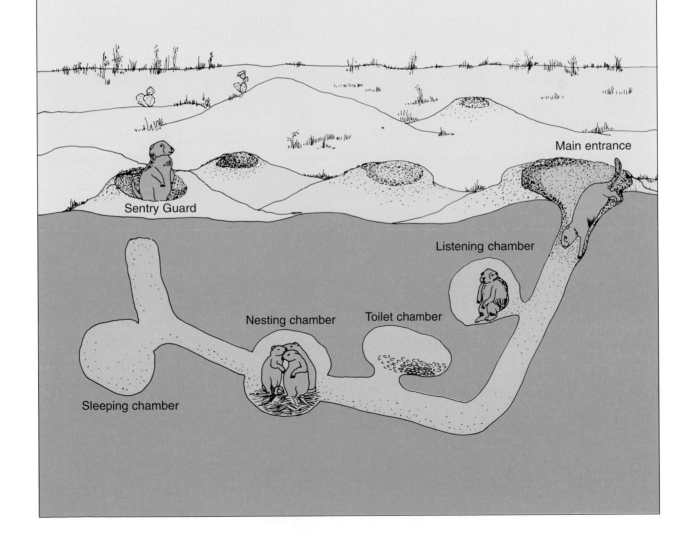

Main entrance

Sentry Guard

Listening chamber

Nesting chamber

Toilet chamber

Sleeping chamber

Prairie dogs recognize each other by touch, grooming, and gestures. Here, black-tailed prairie dogs huddle close together.

is recognized as a prairie dog that belongs to a foreign coterie and is driven away.

The coteries of a prairie-dog town do not engage in organized labor. Because so many animals live in a town, however, some remain on the alert for danger while others feed. These animals have two different alarm calls that all prairie dogs within hearing distance heed. If danger is about to strike, a prairie dog will utter a high-pitched, two-note whistle as it bolts for its burrow. On hearing the whistle, the other prairie dogs immediately rush to their burrows.

When a prairie dog senses danger but cannot confirm it, the creature begins to bark. Nearby prairie dogs will sit up and begin sniffing the air and looking for an enemy. The more frightened the prairie dogs become, the louder and more

44

Rattlesnakes often inhabit abandoned burrows and pose a threat to young prairie dogs.

frequent their barking will be. Once they know for certain that they are in peril, they head for cover. When a prairie dog senses that the danger has passed, it emits a musical whistle that is picked up by other animals throughout the area.

More than 130 species of animals, including prairie-dog predators, live in and around prairie-dog towns. Abandoned prairie-dog burrows are used by the burrowing owl as living and nesting sites. These owls sometimes eat very young prairie dogs, and prairie dogs occasionally eat owl eggs. Rattlesnakes, which also sometimes eat young prairie dogs, may inhabit old burrows. One animal in particular is linked to prairie-dog towns because they are its only home. This creature is a member of the weasel family called the black-footed ferret, which has become extinct in the wild. Scientists have released ferrets born of captive adults in the hope that they will reestablish the species on the western grasslands.

Burrowing Around the World

Several small rodents that are found on other grasslands around the world have lifestyles similar to that of the prairie dog. Susliks live on the steppes of Europe and Asia, where they hibernate in burrows for several months during winter. In some areas of the steppe, they are the most numerous mammals.

The lesser five-toed jerboa, which outwardly resembles a kangaroo rat, lives on the dry steppes of Central Asia and North Africa. It is a small animal but builds long burrows, usually more than 20 feet (6 meters) long. During the day, the jerboas remain within the burrow, after sealing the entrance with earth. At night they come out to feed on juicy plant material.

The plains vizcacha of South America, which can weigh 15 pounds (7 kilograms), looks like a huge guinea pig. It digs large systems of burrows, some of which are used by generations of animals for centuries. A vizcacha burrow system may have up to thirty entrances and cover more than 200 square feet (19 square meters). An entrance may be more than

THE FERRET RETURNS

Late in 1991, biologists from the United States Fish & Wildlife Service and the Wyoming Game and Fish Department began releasing black-footed ferrets into prairie-dog habitat on Wyoming grasslands. These ferrets were the young of animals that had earlier been captured from the wild.

If the ferrets succeed in repopulating the wild, it will be a remarkable achievement. Ferrets depend on prairie dogs for food. They eat few other things. As the West was developed, the population of 25 million prairie dogs, like that of the bison, was greatly reduced. Farms were created on prairie-dog habitat. Ranchers who wanted to protect their means of earning a living killed the prairie dogs because their burrows presented a danger to cattle. They also thought, mistakenly, that prairie dogs competed with their cattle for grass. Today, only about 5 percent of the original prairie-dog habitat remains.

Black-footed ferrets are being raised in captivity and released into their natural habitat.

3 feet (0.9 meter) deep and lead to a chamber that, in turn, branches out into a network of tunnels. Vizcachas spend the day in their burrows, leaving at night to feed on grasses, seeds, and other plant material.

Another South American rodent that burrows under the grasslands, including those of the altiplano, is the tuco-tuco, which resembles a miniature beaver without the paddle-shaped tail. They dig tunnels about 50 feet (15 meters) long—side tunnels branch out in different directions. Like prairie-dog burrows, tuco-tuco tunnels have chambers for storing food and a nursery for the young. The entrances of a burrow are often covered with earth, which the animals remove to let in cool air when the temperature rises.

When different species in different parts of the world develop a similar appearance and lifestyle, it is almost always because they live in similar environments. Many of the burrowing rodents of the grasslands are dramatic examples of this fact of nature.

Many other animals of the grasslands inhabit burrows. Several species of armadillos are found on the grasslands of South America. One is the giant armadillo, which may weigh more than 100 pounds (45 kilograms). Giant-armadillo burrows are often found in termite mounds. Termites are a major source of food for this animal, so it makes its home right where its food supply is. Giant armadillos also eat ants and other insects, spiders, worms, snakes, and sometimes the carcasses of dead animals.

About the size of the giant armadillo and also a termite eater is the aardvark. *Aardvark* is a South African Boer word that means "earth pig," and, with its flat snout and humped body, the aardvark does indeed resemble a pig. Aardvarks have heavy, powerful claws on their forefeet with which they dig their burrows. Aardvarks can dig into soil very quickly, and they easily scoop up ground that is hard to penetrate even with a shovel.

When it digs, the aardvark shovels earth backward under its body as it rests on its hind feet and tail. As the earth piles up, the aardvark pushes it aside with its hind feet, while its forefeet continue to scoop out the soil.

Aardvarks dig a number of different burrows for use at the same time. When surprised in the open, they churn into the ground and disappear. Burrows that an aardvark uses for a while and then abandons are about 10 feet (3 meters) long, with a chamber at the end that is big enough to allow the animal to turn around. Permanent burrows are about 40 feet (12 meters) long and have several entrance holes.

The aardvark lives in the grasslands of South Africa.

49

The badger of North America is another fast digger. Like the aardvark, it has large front claws with which it excavates earth. Badgers burrow more than 10 feet (3 meters) below the prairie and create tunnels about 30 feet (9 meters) long. Digging also helps the badger get to its food supply—ground squirrels and prairie dogs whose burrows offer them little protection from this small but fierce hunter.

A badger uses its sharp claws to dig into the earth.

THE DUCK FACTORY

Prairie wetlands in the upper Midwest and the West were created by Ice Age glaciers, and they are critical to the maintenance of waterfowl populations in North America. These wetlands form around shallow, circular ponds called potholes, the basins of which were scraped out of the earth by the receding glaciers 10,000 years ago. The huge Prairie Pothole Region is an area of about 192 million acres (78 million hectares) in North Dakota, South Dakota, Iowa, Minnesota, and Montana. It produces about half of the waterfowl hatched annually in the United States, although it covers only about 10 percent of the area in which ducks and geese breed.

Prairie potholes are used as breeding and rest areas by many different birds.

Pothole ponds and marshes are key resting places for migrating waterfowl and major breeding areas for several species, especially mallards, redheads, and northern shovelers. North of the Pothole Region, marshes and lakes on the prairies of Canada's Manitoba, Saskatchewan, and Alberta provinces are a key area for mallards and pintail ducks, as well as many types of shorebirds and wading birds.

The rivers, marshes, and lakes of prairies in such states as Texas, Oklahoma, Nebraska, Kansas, and New Mexico are stopover points and wintering grounds for millions of waterfowl and other aquatic birds on their spring and fall migrations. One such bird is the magnificent sandhill crane, which is up to 6 feet (2 meters) high and can have a 9-foot (3-meter) wingspan. It breeds in Canada and Alaska and winters in Texas, New Mexico, and Mexico. In the spring, 500,000 sandhill cranes stop, rest, and feed along the brown waters of the Platte River in Nebraska as they prepare to continue their northward migration. They feed in the marshes and wet prairies that border the river. About a million other birds join them. The prairies are vital not only to the survival of animals that depend in one way or another on grasses, but also to that of aquatic birds.

5

The Future of Grasslands

During the 1930s, drought gripped vast areas of Texas, Kansas, Oklahoma, Colorado, and New Mexico. Farmland and ranges used by livestock turned dusty. Great windstorms carried the dust into the air. Swirling clouds of dust buried the land and even buildings. Throughout the area, ranches and farms failed. People lost their livelihoods, and many were forced to leave their homes. This disaster, which created the region known as the Dust Bowl, was not entirely to be blamed on the weather. For generations, farmers had allowed their livestock to over-graze the pastures and had not given the soil a chance to rest between crops. Vegetation on the grasslands died, and the soil on the farms was exhausted. Exposed to the wind, and no

Opposite:
Strip mining,
shown here
in Montana,
often destroys
grasslands and
established
animal habitats.

longer held in place by the roots of grasses, the once-rich soil turned to dust and was swept away.

Grasslands used for agriculture produce more food for people than does any other biome. This type of cultivation has resulted in the loss of immense areas of natural grassland and the wildlife that inhabited it. There is no question that the production of food is necessary, but so is the preservation of the wild grasslands. In many parts of the world, especially in the developing countries, human populations are skyrocketing. Yet as the need for food increases, poor agricultural practices are destroying both natural grasslands and rangelands. In other places, industry, such as mining, also threatens the natural balance of life in the grasslands.

Advancing Deserts

Throughout much of Africa, what was once fertile grassland is turning to desert—and not the kind of desert in which balanced desert communities live. In these deserts, grasses cannot survive and desert plants are not established—in other words, it is desolate land. The chief culprit is overgrazing by livestock, which prevents plants from growing and eventually damages their roots. Part of the reason for overgrazing is that increasing numbers of people place more animals on the land, until the pressure is more than the vegetation can bear. Some governments have encouraged farming in grasslands too dry to support it. People also use vegetation for firewood. When drought strikes, as it has several times in recent years throughout much of Africa's grasslands, the damaged vegetation dies. However, in their attempt to produce more food, people continue to farm the same exhausted areas. This only does more harm.

The Vanishing Prairie

Although agricultural practices in North America have been vastly improved since the droughts of the 1930s, the real damage to the natural prairie was done long ago. Of the

250 million acres (101 million hectares) of tallgrass prairie that once existed, only 5 percent remains. Much of the land that is left is still overworked, either by farming or by ranching.

Overgrazing has also threatened the grasslands of South America and Australia. In South America, cattle have left little food for rheas. These big birds have dwindled greatly in number and, in parts of Australia, compete with kangaroos for grass.

An aerial view of New South Wales, Australia, shows the effects of over-grazed land.

THE MASAI—PEOPLE OF THE GRASSLANDS

While learning to survive in their changing environment, the Masai of Africa have managed to maintain their traditional ways for centuries.

The Masai of southern Kenya and northern Tanzania have since ancient times been linked to the grasslands for their survival. They depend on their livestock, chiefly cattle but also sheep and goats. The Masai consider their cattle their wealth. They also believe that God gave them all the cows in the world, which is why they have believed that it was their right to raid the herds of other tribes. The chief food of these people is milk products. Their cattle are seldom used for meat. In fact, the Masai eat little meat; what meat they do consume comes from sheep and goats.

The Masai are famed warriors, and fiercely independent. Many of them do not recognize the national border between Kenya and Tanzania and move across it at will. Increasingly, however, they are bowing to the pressures of civilization, and large numbers of them have settled down and turned to farming.

Traditionally, the Masai were nomads. Like the wild herds of the African grasslands, the Masai livestock was continually moved in pursuit of the rains. Their encampments were basic. A circular wall of thornbushes protected their livestock and huts.

The herds of the Masai compete for grass with antelope and other grazers. The Masai often try to bring their cattle to grazing areas before their wild competitors arrive. This sometimes drives the wild creatures from the best grazing areas. However, wild herds can spread out over wider areas and can feed all night long. The Masai's cattle are returned to camp before dark.

When the Masai were completely free-roaming, their cattle were not in one place long enough to overgraze the vegetation. Because their movements today are somewhat restricted, their herds sometimes concentrate in one area for too long and overgrazing results. Farms and ranches now occupy much of the land once used by both the Masai and the wildlife. This means that both put more pressure on the open land that remains. Some of the Masai have agreed to keep their herds out of wildlife preserves in return for a share of the money brought in by tourists who come to view the animals. It is an arrangement that works out well for the wildlife and the Masai.

Conserving Grasslands

Saving the grasslands that remain on Earth is not an easy task, but conservationists are working hard at it. Soil can be conserved by giving it a rest from cropping and grazing. Groves of trees between fields of grain help prevent erosion of soil by the wind. Still, many conservationists believe that the grasslands, like the other biomes, will not survive in those parts of the world where human numbers are expanding too rapidly for resources to support them unless population control is observed.

A significant stumbling block to conservation in developing countries is that it is difficult to persuade a starving farmer to leave a grassland alone. In some countries, such as Kenya and Tanzania, large areas of grassland have been set aside as parks and preserves. People living outside these areas naturally want to use them to make a living. One way around the problem is to involve local people in the operation of the parks— and to profit from fees paid by visitors to the parks. This technique is working in both Kenya and Tanzania and in many other places, and it has encouraged people to conserve both habitat and the wildlife that uses it.

The Serengeti National Park in Tanzania and an adjoining preserve in Kenya safeguard most of the area used by the great wildebeest herds. Both countries are poor and depend on funds and other support from developed nations to maintain this great natural treasure.

Several national parks in the United States and Canada also protect grasslands and grasslands wildlife. About 15,000 bison inhabit Wood Buffalo National Park in Canada, which lies where the prairies meet the northern coniferous forest, or taiga. Custer State Park in South Dakota has 1,500 bison; Badlands National Park, also in South Dakota, has 500; North Dakota's Theodore Roosevelt National Park has 300; and the National Bison Range in Montana has up to 500 of the big, shaggy creatures.

A herd of bison travels across the National Bison Range in Montana. The range was established in order to protect the bison and other grassland species.

A major achievement in grasslands preservation occurred in June 1994, when the National Park Trust, a conservation organization in Washington, D.C., purchased a 10,894-acre (4,412-hectare) ranch in the Flint Hills. One of the largest tracts of land is the Flint Hills of Kansas, an area of tallgrass about 50 miles (80 kilometers) long that has not been farmed, because of thin topsoil, but has been used by ranchers for their cattle. Conservationists have long hoped to see a national park in the Flint Hills and view the purchase of the ranch as a first step. However, many people who live in the area, including some ranchers, oppose turning the ranch into a park because they want the area open to grazing. If the government decides to create a park, it could force people who own land in the Flint Hills to sell it to the National Park Service, a possibility that many local property owners fear. As it does in so many other conservation issues, politics plays a role in the future of wild habitats. Conserving the grasslands and all of the other types of habitat is not just a matter of ecological science. The needs of people—whether they are poor farmers in Africa or ranchers in Kansas—must be considered if natural areas are to survive.

BLACK THUNDER

Strip, or surface, mining is the process by which huge amounts of topsoil and earth are removed from the land to expose beds of minerals, such as coal. This technique has destroyed significant areas of grassland in the western United States. However, with considerable effort, grasslands can be restored after mining ends. This has happened at Black Thunder Mine, which is operated in northeastern Wyoming.

Three hundred acres (122 hectares) of grassland are mined a year, with no more than 1,000 acres (405 hectares) remaining disturbed at any one time. Soil is removed and stockpiled the first year; coal is mined the second year; and the restoration of grassland begins the third year. Government regulations stipulate that the restored land must have the same contours, drainage patterns, and plant communities that it had prior to the mining. The ecology staff of Black Thunder Mine not only complied with the regulations but went well beyond what was required by creating a community of prairie vegetation much closer to natural prairie than the grazing land that was there before the earth was stripped away.

Wildlife came to the restored land, sometimes in greater numbers than they had before. Mule deer, elk, and pronghorn began to feed in the area, and, according to a survey done in 1989, six species of hawks and eagles were nesting there. These included golden eagles and the rare ferruginous roughleg hawk. Six pairs of these grassland hawks bred at Black Thunder in 1989, including two on nests that had been removed because they got in the way of the mining. After mining stopped, the nests were replaced, one on a platform the other on a rock pile that the mine staff had reconstructed atop a ridge.

Government regulations demand that strip-mining companies build reservoirs to collect the sediment carried away by water during the digging of coal. In 1986, a 240-acre (97-hectare) reservoir was built at Black Thunder Mine. It was constructed in such a way as to attract aquatic birds. The shoreline was carved in an irregular fashion, and islands were created. Waterbirds that made use of the lake include eared grebes, redhead ducks, willets, white pelicans, and black terns.

Ferruginous hawk nestlings rest atop a nest that was reconstructed by the Black Thunder Mine staff.

Glossary

altiplano A treeless grassland plateau, or high plain, located in the central Andes Mountains.

badlands A prairie grasslands landscape characterized by dramatically sculptured rocks and buttes and sparse vegetation.

blade The leaf of a grass.

burrow A hole or excavation in the ground made by an animal.

butte An isolated hill or mountain with steep sides, having a smaller summit area than a mesa.

carnivore Any animal that feeds primarily on flesh.

carpel The female part of a flower.

conservation The protection and preservation of the natural environment.

coterie The family group of the black-tailed prairie dog.

drought A period of dryness.

erosion To wear away by the action of water, wind, or glacial ice.

herbivore Any animal that feeds primarily on plants.

migrate To move seasonally or periodically from one region or climate to another.

overgraze To allow animals to graze (eat vegetation) to the point that the area's vegetation is depleted.

pampas Treeless grasslands in the temperate South American region east of the Andes Mountains.

pluvial A prolonged period of abundant rain.

prairie Treeless grasslands found in North America.

predator An animal that hunts other animals in order to eat them.

prey An animal that is eaten by another animal.

pride A family group of lions.

rhizome An underground shoot by which a plant can reproduce.

root The part of a plant body, usually underground, that serves to anchor and support the plant as well as to store nutrients.

savanna A grassland with scattered trees, particularly in Africa and parts of South America.

scavenger An animal that feeds on carcasses of dead animals.

Serengeti Plain A grasslands region of Africa.

stamen The male part of a flower.

steppes Grasslands in Eurasia.

stigma A sticky structure on a carpel to which pollen attaches.

For Further Reading

Collinson, Alan. *Grasslands.* New York: Dillon Press, 1992.

Kaplan, Elizabeth. *Temperate Forest.* New York: Marshall Cavendish, 1996.

_____. *Taiga.* New York: Marshall Cavendish, 1996.

Lambert, David. *Grasslands.* Morristown, NJ: Silver Burdett Press, 1989.

Patent, Dorothy H. *Prairie Dogs.* Boston: Clarion, 1993.

Ricciuti, Edward R. *Birds.* Woodbridge, CT: Blackbirch Press, 1993.

_____. *Chaparral.* New York: Marshall Cavendish, 1996.

_____. *Rainforest.* New York: Marshall Cavendish, 1996.

Tesar, Jenny. *Endangered Habitats.* New York: Facts On File, 1992.

_____. *Mammals.* Woodbridge, CT: Blackbirch Press, 1993.

Index

Acknowledgments and Photo Credits
Cover: ©Bill Bachman/Photo Researchers, Inc.; p. 6: ©Lew Eatherton/Photo Researchers, Inc.; pp.
8–9: ©Michael Fairchild/Peter Arnold, Inc.; p. 10: ©Bates Littlehales/Earth Scenes; p. 11: ©Calvin
Larsen/Photo Researchers, Inc.; p. 14: ©Lawrence E. Naylor/Photo Researchers, Inc.; p. 16: ©Mitch
Reardon/Photo Researchers, Inc.; p. 17: ©S. J. Krasemann/Peter Arnold, Inc.; p. 18: ©Patti
Murray/Earth Scenes; p. 21: ©C. Prescott-Allen/Animals Animals; p. 21: ©Tim Davis/Photo
Researchers, Inc.; p. 23: ©Ted Levin/Earth Scenes; p. 23: ©John Lemker/Earth Scenes; p. 27: ©Anup
Shah/Animals Animals; pp. 28, 39: ©Joe McDonald/Animals Animals; p. 30: ©Gregory G.
Dimijian/Photo Researchers, Inc.; p. 52: ©David C. Fritts/Animals Animals; p. 33: ©Michael
Dick/Animals Animals; pp. 34, 58: ©François Gohier/Photo Researchers, Inc.; p. 35: ©John
Cancalosi/Peter Arnold, Inc.; p. 36: ©Sven Lindblad/NAS/Photo Researchers, Inc.; p. 38: ©Fran
Allan/Animals Animals; pp. 40, 47: ©Steve Kaufman/Peter Arnold, Inc.; p. 44: ©Willard Luce/Animals
Animals; p. 45: ©Michael Fogden/Animals Animals; p. 49: ©Patti Murray/Animals Animals; p. 50: ©E.
R. Degginger/Animals Animals; p. 51: ©Brian Milne/Earth Scenes; p. 52: ©Barbara Pfeffer/Peter
Arnold, Inc.; p. 55: ©C. Prescott-Allen/Earth Scenes; p. 56: ©Bildarchiv Okapia/Photo Researchers,
Inc.; p. 59: ©L. W. Richardson/Photo Researchers, Inc.
Artwork and graphics by Blackbirch Graphics, Inc.